Paul Tyner

Bodily Immortality

Paul Tyner

Bodily Immortality

ISBN/EAN: 9783743441828

Manufactured in Europe, USA, Canada, Australia, Japa

Cover: Foto ©berggeist007 / pixelio.de

Manufactured and distributed by brebook publishing software
(www.brebook.com)

Paul Tyner

Bodily Immortality

BODILY IMMORTALITY.

BODILY IMMORTALITY.*

"Truth is within ourselves; it takes no rise
From outward things, whate'er you may believe:
There is an inmost center in us all,
Where truth abides in fullness:—"
—Robert Browning.

TURN with me, if you will, to this "inmost center in us all;" let me ask you to lay aside, for this hour, all prejudices and preconceptions formed in your minds by the traditions and teachings that lead men to live so largely in outwardness, rather than inwardness, in form rather than substance, in letter and not in spirit. Whether your materialism be secular or theological; however sufficient or conclusive may seem to you the knowledge or belief

*—An address delivered before the Congress of Truth, Unity Church, Denver, Colorado, June 1, 1896.

that for the moment limits your horizon
and so colors your attitude towards
every new presentation of truth, the
fullness and clearness of the word I
have to give you to-day will be helped
in the giving and in the receiving, if at
the start our minds are joined in the
single thought of desiring only truth,
whatever form it may take.

An innermost implies an outermost;
and inner and outer, center and circumference, are *one*. If I take you, first
of all, into your inner selves, it is only
that you may the better understand the
outer self and that this *oneness* of inner
and outer may be emphasized.

Man is enabled to look inward because he can look outward. The spiritual vision is developed with the physical vision. Imagination would be impossible in a blind race. What we call
the two worlds are only two sides of
the one world. The curtain that di-
vides the visible from the invisible is
lowered by the imagination; it may
be raised by the same power. Every

spiritual truth grasped by the spiritual vision exists in outwardness, as well as inwardness, and its material manifestation to the physical perception only awaits *recognition* on the physical plane. "The kingdom of Heaven is within you,"—but when you have found it nothing is without you. There is heaven everywhere.

If it be true that "Man is a spiritual being and the spirit is the man," then *all* the man, outer and inner, is spirit. The real must be real on both planes. Whatever is true in the great is true in the little. That which we call the finite is an expression of the infinite; none the less, the infinite dwells in every atom of this "finite." God's expression must be as infinite as is his essence. The limitations of form are not fixed, but expansive and expanding ever. Time and space are pushed constantly into the unending and limitless. Man's only true relation to God—God's absolute relation to man—is *oneness*. An *infinite* God outside of His expression in man and man's world, would mean God outside of Himself—that is to say, a God that is inconceivable. "God is everywhere," says the Church catechism, recognizing the logical necessity of omnipresence in deity; "God knows all things, sees all things, can do all things, is without beginning and without end." Does it not follow that wherever in this "everywhere" that God is, *there* is omniscience and omnipotence? Once man comes into consciousness of his oneness with Omnipresent, Omniscient and Omnipotent God, what can he lack to exercise and express the God in him? If God is immortal, man is immortal; and if man is immortal in spirit, and spirit is all, the man is immortal, and may—nay, he *must*, express his immortality in all his being,—in his outerness as well as in his inwardness, or rather, in *both* his outerness and his innerness made one.

Realities are both subjective and objective. The spiritualist who deems the material universe as less real than the spiritual world, makes the same mistake as does the materialist to whom the outer appearance only is real, all else unreal. Both spiritual and material advance have for centuries been hampered and hindered by this error which we owe, in great degree, to the religious teaching that still gives it large, though rapidly lessening, place. We are learning that dogmatism is as much a mistake in science as in religion.

> "New occasions teach new duties,
>   Times makes ancient good uncouth.
>   He must upward still and onward,
>   Who would be abreast of truth."

Between Spirit and Matter there is no real barrier. The biological phenomena, which so perplex our modern scientists, are explicable on the comparatively simple hypothesis of controlling mind. The Kabbalistic theory that "Spirit informs all matter," finds implication, if not acceptance, in Bishop Berkeley's teaching that: "There is only one substance and that substance is spirit." It is not the less implied in modern physical philosophy. Prof. Winchell, in his "World Life," boldly declares:

"But one system of matter pervades the immense spaces of the visible universe, and all the recognized chemical elements will one day be found but modifications of a single material element."

Dr. Crookes' remarkable experiments on so-called "radiant matter" and the more recent marvels of Dr. Roentgen's X-rays, are best understood on the hypothesis of the *homogeneity* of matter and the *continuity* of the states of matter.

Manifestly any rational conclusion as to the homogeneity of the primal substance must involve the *spirituality* of that substance. Matter which thinks is clearly *not* the matter of the dualists. The contention that life and thought are merely properties of matter presupposes a matter differing only in name from what we call "spirit." All forms of matter, are but bound up energy. That form of energy we call "thought" is a universal solvent. Thought, consciously or unconsciously exercised, is constantly changing the condition and the form of the human body. Control of this thought-force thus means the power at every moment to determine whether the body shall show forth life in fuller or in lesser degree. It rests with man to say whether his soul shall be housed in a stately mansion of ever-growing splendor and beauty, or in a hovel of his own building—a hovel at last ruined and abandoned to decay.

Knowledge, according to Plato, is gained by three means: by *Perception*, whence is derived the knowledge of physical facts and happenings, which when classified, constitutes our Science; *Reason*, arising from the comparison of known facts, the inductive apprehension of causes in relation thereto, and the conception of laws inclusive of these, whence Philosophy; and *Intuition*, a direct cognition of spiritual things by the inversion or turning in of the mind, followed by the presentation of such spiritual facts in terms of our normal consciousness under figures or "corre_

pondences" of the external world, and
the application of these truths to the
life of man, whence Religion.

To these three means of obtaining
knowledge, we have, in our day, added
a fourth; which finds large place in
the best modern schemes or systems of
education, and which we may call *Expression*—Art's larger name. Seeking,
through art, to express or externalize
in form and color, in sound or action,
the world of effects, the world of causes
and the world of principles, as they are
imaged in our minds, we find ourselves
—at every stroke of the brush, every
touch on keys of piano-forte or organ,
every chip of chisel or blow of hammer,
—coming into clearer and surer perception of truth.   The best of all ways to
learn to swim is to—swim.   The advice
of every really great teacher in art to
the aspiring student is: *"Paint, paint,
paint!"*   All that masters in literature
can tell beginners with the pen, is:
"Write!"   No amount of reading or
theoretical instruction can make an

orator of the man who does not speak. And it seems to me that the best way to learn a thing is to do it; simply because, in the doing and the result of our doing, we are compelled to unite all other modes of learning: to perceive, to reason, to feel. "He who *doeth* the will of the Father, shall know the doctrine."

It will help us, I think, to recognize that "perception," Plato's first means of knowledge, really covers all methods of obtaining knowledge. It is perception, after all, that is enlarged, deepened, clarified and verified by Reason, Intuition and Expression. True religion is identical in scope with true philosophy, and both should be based on *Knowledge*—on knowledge gained not through any one mode of perception alone; but on knowledge verified through all modes of perception. The mental arrogance of the Rationalist, who refuses to consider as facts any phenomena not accounted for (as he imagines) by the known laws of the natural world, is

hardly less reasonable than the arrogance of the Religionist, who refuses to
admit the demonstrated facts of physical science into any relation with the
facts of the spiritual world,—as *he* imagines them.

The development on the intuitional
side of the perception of an ideal Christ
and the imaging of that ideal, led the
author, in course of time, into the
perception, through the physical senses,
of a concrete, personified embodiment
of that ideal, in a living man of flesh
and blood!   Not only does this perception make his ideal a larger, grander,
more beautiful and more vivid reality,
in the opening and illumination of consciousness as to the *meaning* for all
mankind of this fact; it also forces him
to recognize that every truth evidenced
in an individual is writ large in nations,
in humanity,—in Nature herself.

The resurrection of Jesus, the Chris-
ed or spirit-baptized and anointed
man, in the body of flesh and blood,
and his continued personal existence in

that living body of flesh and blood, among men on earth—this is the basic fact on which rests the truth of bodily immortality. And this fact has been made known to the author by perception on the outer and on the inner planes. Because the Christ lives in the body, he does not live less but more in the spirit generated in and radiating from that personality to all men— through all the world and through all worlds. Yet, this truth is perhaps not readily nor easily reconciled with what, to many of us, is a high and holy conception of the nature of God and man. Here we have the result of the repetition, through so many ages, of the manifestation of *Death* in the body—the great error, the great negation the great lie!

Death in the personal body of so many of earth's greatest and wisest seems to us to make greatness and wisdom impossible in living persons. Because Plato and Shakespeare and Dante and Virgil live in their works so largely,

and because they have not remained
with us in the bodies of flesh associated
with their personalities—in which
and by which, and through which, they
gave us their works—the very thought
of their *living* spirits seems to suggest
their *dead* bodies. To a large extent
we make of these living works mauso-
leums, sepulchres, monuments, to mark
and memorialize not the life, but the
death, of our great. The world goes
forth in brave panoply, with blare of
trumpet and beat of drum, in a Holy
Crusade, to rescue from the Infidel—the
empty tomb of the Saviour!

As well separate the light of the sun
from the sun itself, as separate the life
of a man from that man's embodied
personality in flesh and blood. If the
concrete, material structure and the
distinct form, life and motion of the
great orb of day should cease, the
energy that lights and warms and vitali-
zes the earth would cease very soon
also.

Jesus in his personal body of flesh

and blood, is the "Sun of God," the center generating and radiating God's essence, God's spirit, the light of truth and warmth of love, to all men. He is also, in spirit and flesh, the "Sun of Man," receiving, focusing and reflecting back to God, in reciprocal vibration, all the light and love generated by humanity's evolution—its movement upward and forward through all the ages.

Here, may be briefly stated the conception of the nature of God and of Man, which conscious knowledge of Jesus, the Christ's, personal existence in flesh and blood, here and now, brings to the author. The old "ashes to ashes" idea of the body, which dooms all flesh to the grave, was for years as dear to the author as it could be to any one. His present idea seems to him a larger and grander one. Whether or not it will seem so to others, he cannot tell.

GOD is universal and infinite spirit, essence or energy without beginning and without end. Because God is this, by the law of His being, the transmutation of this energy into work, *i. e.:* His *expression* or manifestation, is *equally* without beginning and without end, because only *infinite progression*, in finite expression (all expression being in a sense finite) can manifest infinite perfection in essence.

MAN, in the large sense, includes the universe. Nature is man writ large, as the history of nations is the history of individuals writ large. Man *is* the universe; it is of him as he is of it. *Man* is the universe, and the universe is man; as much in outerness as in innerness; on the visible, no less than on the invisible side. Man's *physical body* is the microcosm of the visible universe, (of all he senses consciously or unconsciously), as his *soul* is the macrocosm of that visible universe, in that it *is* itself the invisible universe, of which the visible universe is but the microcosm. Man, as the latest and highest, the most perfectly organized manifestation of that infinite and eternal energy from which all things proceed, contains and sums up in himself all the lesser manifestations that preceded his appearance. In his bodily substance and organization—that is in his body of flesh and bones—he represents that *combination* of elements which, as familiar experiments in chemistry show us, holds all the qualities and powers of every element it contains, and holds also an *added* and superior—a dominating—quality and power, not contained in any one element in itself, but arising from their combination. This added and superior quality and power it is which we recognize distinctively as *Man*. By reason of this, by reason of what he is, man dominates all; is given dominion over earth, sun, moon and stars; over seas and mountains, winds and waves; over every green thing that grows, over all the beasts of the field, all creeping things, all living things that swim in the sea, or fly in the air. This man is, just to start with, in the primitive savage, the first Adam. In the second Adam, the Divine Man, the Christ, he is the still higher development of a new combination: that of men—mankind—he is now racially and individually that Christ— that manifestation of the divine, in the human, of which Jesus is the tree and we are the branches.

Daily and hourly the meaning which the recognition of this fact of Christ's living presence on earth has for all men expands in my view. In the immortal God manifested in *immortal man*, we have at last the long sought basis for a perfect union of all the various branches of the Universal Church. In his embodiment of all that is highest and best in the teachings and aspirations of all religions, Christ furnishes common meeting ground for Buddhist and Brahman, Moslem and Jew. In the added truth, the crowning truth, which his continuous life in the flesh now gives to the race we have substantial reason for the preaching of his gospel "to all nations," and especially to those whose own great teachers have given them ethical codes and ideals to which Christ-ianity, *minus the living Christ*, could really add nothing.

In the *visible manifestation* of his one-ness with God, *through* oneness with man, in absolute love—"other-worldli-ness" will be banished and all the grand forces of religion will be directed to lifting this life and this world into what they should be and what they will be; —to bringing the kingdom of heaven on earth, to giving the City of God, de-scending out of Heaven, earthly place and power. Coming not to destroy, but to fulfill the law (the measure of truth) given before his coming not only to the Jews, but also to all the other great races of men, he has no quarrel with any existing religion, on its posi-tive side, only fulfillment, only realiza-tion, only love!

In philosophy, by the supreme test of his own personal life and its influence upon the welfare of the race, he brings reconciliation between Idealism and Utilitarianism. He brings peace and

order into our social conflict and unrest,
by his personal demonstration of the
truth of social solidarity in a more literal and absolute sense than most so-
cialists dream of.   His life affords irre-
futable evidence that, for weal or woe,
for better or worse, human society is
absolutely one grand living organism,
with closely interrelated structure and
function, as actually as is the body of
the individual man.   He shows us that
the health and happinness of every unit
of society, (which means above all the
healthy, constant and harmonious *acti-
vity* of each unit in its relation to every
other unit in its proper place in the organism), are vitally concerned in the
health and happiness of society as a
whole.   Or rather, since the whole
comes before the part,—that the health
and happiness, which means the integ-
rity, the beauty, the freedom, the vigor
and the power,—of the collectivity, is
vitally essential to the health and happiness of every individual composing it.

Christ meant what he said ever and

always; not allegorically nor figuratively, nor fancifully,—but actually and
truly, and literally, in the plain, every-
day meaning of the words, as a
child would understand them. When
he spoke figuratively or in parables, he
took particular care to say so. He
took pains to avoid misleading his
hearers. When, for one thing, he said
"Inasmuch as ye have not done it unto
the least of these, ye have not done it
unto me," he meant that onenesss with
God required a recognition of the fact
that while a single child remains neglected; while she or he lacks all
of opportunity, nurture, care, comfort,
training and education, that the largest
love commanding the largest resources
can give,—then is Christ neglected and
the Father who sent him neglected, the
God in us denied, the Christ in us crucified—and that whatever it is that we
do give these little ones, and so to God,
denying this, is less, much less, than
that LOVE which is the essence of all
real religion As Ruskin well says:

"Anything which makes religion its second object, makes religion no object. He who offers to God a second place, offers Him no place."

The LIVING CHRIST,—living the highest, holiest, happiest and most sublime life that mind of man has been able to conceive, here and now on this mundane sphere, and in the same body of flesh that was nailed to the cross nearly nineteen centuries ago;—the *same* in as true a sense as that intended when we speak of the body that was crucified as the same in which the boy Jesus disputed with the doctors in the Temple;—this living Christ must bring Christendom to a clearer and livelier recognition of the truth that the body in a real and literal sense is "the temple of the living God." This living Christ, in glorified body points plainly to the recognition in every state or national system of education, of the immense importance of giving the fullest consideration to the needs of the body, beginning with the babe before and after

birth (so ordering our social life that
to be "born without sin" shall be the
rule and not the exception), and em-
phasizing clearly and unmistakably the
right and justice, as well as the wise
politics and economics, of the demand,
that, not in the rich man's home alone,
but in the homes of *all* the people, the
body must have fullest and freeest nurture and development—be well and regu-
larly fed, bathed, trained and exercised
in all wise ways.   In other words, He
brings home to us, to every people in
their corporate, communal and national capacity, the truth that Life in any
large and true sense, for nation or individual, requires first of all that every
boy and girl, every man and woman,
in that nation, must share equally and
fully in all opportunity for knowing the
joys of perception and creation, impres-
sion and expression; that if any of us
would follow Christ and have eternal
life, we must consider the bringing
about of an order and arrangement in
government and society which will

secure this equality of opportunity, as the most important and immediate thing in the world to be done; as the thing God wants us to do FIRST.

So, in short space, we shall attain, among other things, to that beauty of the ancient Greeks, which, beginning with recognition and appreciation of the flowing lines of the human body, has left us beauty in architecture and sculpture that has been the inspiration and delight of succeeding ages.

From our living Christ, we shall learn the truth of all this as shown in *his* life, —we shall learn how to come to that higher beauty which answering Socrates prayer "to Pan and all the other gods," shall fill the inward soul and "make the inward and the outward man at one." We shall have outward as well as inward beauty that is true and enduring, when we have attained the power to manifest Eternal Life, as Christ has attained it—through oneness with man in all-embracing *love*.

It is true that we have very generally

acknowledged *in words*,—the power and the wisdom and the beauty of Christ's summary of the law and the prophets. In countless sermons and many hymns, the "new commandment" that he gave us, "That ye love one another," has been held up to the admiration of mankind. But we do not love one another in any true or large sense,—at least we do not show our love. If we *did* love one another, we should not lie to one another, nor swindle one another, calling it "business;" nor would any sane man or woman find enjoyment in possessions and pleasures purchased at the expense of misery, deprivation and suffering to thousands of those "others," whom we are adjured to love. When we do *love one another*, the security of society will be obtained without the terror of the gallows and the guillotine, without the maintenance of standing armies of fighting men to restrain the famishing, and without the condemnation of thousands of those "others" whom we should love, to degradation, to dehumanization —to a living death within prison walls.   As Tolstoi puts it:

"I will be certain that my piece of bread belongs to me only when I know that everyone else has a share, and that no one starves while I eat."

Other-world religion and the saving of the soul *beyond* the grave have had their day.   It is time that we should have a this-world, and this-life, a this-body and this-time —a here-and-now religion.

Through the mistaken belief in Christ's death, or rather through the failure of men to believe in his triumph over death in the resurrection, we have somehow *made*, falsely made, a wide separation between the life here and what men call "the life beyond;" between the soul and the body, between heaven and earth, between time and eternity, between God and Man!   In this error we may find, I think, one great cause for the pre-eminent weakness which distinguishes the Christianity of our time from other of the great religions of the world.   Brahman and Buddhist, Jew and Mahomedan, Parsee and Confucian are all at least *consistent* and sincere in their endeavor to *practice* the teachings of *their* great teachers. "Christians" alone protest that, however beautiful as abstract ideals may be the precepts of Jesus, they are impossible.

The crucified saviour, the entombed saviour, means that in following him we are to expect only sacrifice and loss, persecution, suffering, death: a cross here, a crown in the next world. To preach Christ crucified and not Christ risen is to exalt death over life. It is to ignore the triumph of Christ over Death, —to picture him as vanquished instead of victor.

Of course, the soul of the developed man has instinctively rejected this so-called "religion." Man would not be man, the highest manifestation of God's life, ever advancing, ever rising,—if he did not *prefer* gain to loss, joy to sorrow, pleasure to pain, life to death,— and if he did not want all possible goodness and sweetness and strength, and

power and glory and happiness, here
and now in the life and in the world in
which he finds himself, by God's wise
plan,—rather than in some life to come,
in some other world! Whatever that
after life in that other world may be,
man is compelled by reason and in-
stinct to insist that it be founded on
and developed from, made in and out of
this life in this world. Progress is the
eternal and immutable law. Men may
die; civilizations may decay; races may
fade out and all memory of them pass
into oblivion; but MAN lives on,—man
is immortal—and by reason of his inhe-
rent immortality and in spite of his
failure to recognize it, the world is more
and more.

> . . . . "forevermore
> With grander resurrection that was feigned by At-
>   tila's fierce Huns,
> The soul of Greece conquers the bulk of Persia."

In a full recognition and more perfectly balanced expression of the com-
plementary principles in humanity, typi-
fied in the masculine and the feminine
elements throughout nature, and in the

supreme development of sex in humanity, must be found, I think, the key to immortality in *expression*, as well as immortality in essense; to undying life *in the flesh*, as in the spirit; in *art*, as in the idea it seeks to embody; in the *form*, as in the substance.

Beginning among the ancient Egyptians with a rapturous worship of the feminine principle in Mother Isis as so far beyond all external nature that it was impossible to express her glorious perfection wholly, and worse than useless to express it in any degree, we find Woman—for a while women,—venerated, hailed ruler by divine right, consecrated as inviolate priestess in the Temple, embodied in the Sphinx as incomprehensible but worshipful mystery. A little later, we find women shut up in convents and in harems. She was robbed by degrees, first of her humanness, and, at last, in the Ottoman empire of to-day, as in varying degree throughout the East, she is *denied a soul;* degraded spiritually, *because* degraded *humanly!* God requires not merely *a man*, or some men; but Man, male and female, in largest and best development, for the fuller expression of His thought; that thought which is of His very nature and which when manifested, becomes the God-Man. This divine unfoldment we see perfected, first in the individual, and next in the social organism. Humanity to-day, more than ever before in the world's history, (as a result of social evolution marching with accelerated and ever-lengthening strides since the industrial revolution which ushered in our century), consists not of mere aggregations of individuals; but of organized political units, of men living in orderly relation. Because of the very increase in complexity of structure and function, which distinguishes the higher organism from the lower throughout nature, we find this orderly relation developing greater and greater *interdependence* among nations and societies of men, as between the members of these nations and societies.

Out of the very heterogeneity that is part of our development, and out of the necessary complexity of any arrangement in right relations of the parts to the whole and of the whole to all the parts, we are evolving a homogeneity of feeling greater in intensity and extensity than was ever before known; that is, a need and a capacity for "loving one another" in larger and deeper ways than were possible in the primitive organizations of men. The life of the hermit or recluse—the life of any isolated community, city, village or nation—cannot, in our day, possibly be *human life*, in any true sense of the word. And I think you will agree with me that, not being human, it certainly is not Christian, is not spiritual, is not religious. "If a man loveth not his brother whom he may see, how can he love God whom he does not see?"—that is does not see otherwise than as he sees God in man. The essence of religion is *love*—love in its finest, fullest, largest, truest develop-

ment. "Love to God and love to man"
is to be realized practically only by
Love of God in Man. "Love one another!" is not the command of Jesus
merely, not the mandate of an awful
Jehovah from a burning bush; it is not
the injunction, simply, of ancient or
modern prophet or law-giver. It is the
immutable law of all life throughout
the universe; the law inscribed upon the
open scroll of the heavens, and written
on sea and land, ages ere man had hewn
tables of stone from the mountain side,
or traced his thought in books. It is
the law that holds seas and mountains
in their places; the law that builds suns
and planets and that governs their mar-
velous order and procession. The moment we obey this simple commandment, that we "love one another," that
moment will usher in the millenium.
That moment will begin a reign of
health and harmony, of peace and progress: not only in the social organism,
which represents humanity in the large,
but also and equally in the individual

organism,—in the body of flesh and blood of each and every one of us. And surely that time is at hand.

For nearly two thousand years, we have, blindly and foolishly, darkly and dogmatically, in sorrow and suffering, and with cruelty and bloodshed, been trying to lift up the Christ, to lift up the glorious personality of Jesus of Nazareth, and the principle of Truth embodied in that personality, by SEPARATING him from Man; by debasing and degrading man. At last, we are learning that he can not be truly lifted up, until ALL MEN are lifted with him. We are beginning to understand that, however great and good Jesus was and is; however resplendent and noble, however worshipful and divine, he is all this by reason of *what we are.* Jesus owed the manifestation of divinity in his personality to the divinity contained in the whole race, as truly as the energy manifested in the highest wave of the ocean is the energy derived from all the ocean's mass, to its widest reaches and its profoundest depths. The "glorious orb of day" is glorious, not in itself alone, but because of the reciprocal and related motion, vibration, of all the planets in our solar system, and of all solar systems; of all the sixty millions of stars revealed to us through the telescope, and of the unknown millions yet to be revealed. The height of a mountain is the height of its topmost peak, whether the stones about the base are conscious of the fact or not. You cannot lower the base without lowering the summit. If oneness with God means anything, it means oneness with man! If oneness with man means anything, it means that wherever a man, woman or child is burdened or oppressed, I am burdened and oppressed. It means that while a single human soul is starved physically, mentally or spiritually, I am starved. It means the suffering of the poor is my suffering; the blindness of the blind my blindness; the wrongs of the worker my wrongs; the grief of the broken-hearted my grief; the bonds of the slave my bonds. Oneness with God and Man means all this, and, BECAUSE it means all this, it means much more. It means oneness with *humanity's power of conquest* over all conditions negative to perpetual growth in strength and power, beauty and happiness. It means command of unlimited supply for every need. It means oneness with the Christ who shall give sight to my eyes, heal my sickness, cleanse me of my leprosy, feed my hunger, clothe my nakedness, loose my bonds, comfort my broken heart, heal my wounds, and raise me from the dead, triumphant over the last enemy and so triumphant over ALL.

I AM that which I see in my brother; that which I love in him.

"Whate'er thou lovest, man, that too become thou
    must;
  God, if thou lovest God; dust, if thou lovest dust!"

GOD'S oneness is manifest most perfectly in the ONENESS OF MAN; and His true worship consists in the WORK which testifies to recognition of that oneness. It is not the SEPARATION of the outer

from the inner; of form from substance;
sense from soul; man from woman; the
"practical" life from the "spiritual"
life; the concrete from the abstract;
the actual from the ideal; God from
our neighbor, that is to be emphasized
in true religion, in true philosophy, in
true politics and economics. It is not
their separation from each other, but
their RELATION to each other,—their
unity. It is their *equal holiness in equal
recognition*, that is to be emphasized.
"The life is more than the meat and
the body than raiment"—but that is
only a reason for constantly larger,
clearer, purer manifestation in "meat
and raiment" (as in "the body"), of
that LIFE which, in itself, being always
MORE, (not different or a thing apart),
always requires ever finer and finer food
for its nutrition; ever more and more
beautiful vesture for its outer adorn-
ment. The *profession* and preaching of
Righteousness, Justice, Love, is of no
avail, if our practice, collectively,—that
is HUMANLY —be allowed to fall behind

our profession and preaching. To praise God with both hands and lips is the need of our time; the only HONEST praise.

There is a reason for everything, and I think we may find the sufficient reason for the failure of our present day conception of practical religion (which I may be permitted to point out parenthetically, finds perhaps its best exposition in the activities of the Salvation Army at one end and those of the Society for Ethical Culture at the other), —we may find this failure accounted for in the illusory, I may even say, unreasonable, limits placed on the personal, individual life in the flesh.

Surely it is but a half-hearted allegiance that a man gives to the demands of THIS life in THIS world, if he regards it as only a "fleeting show" a passage to another life in another world, more real and more enduring! Works without faith are little better than faith without works. And what faith, what importance, can a man yield to the service of a country in which he regards himself not as a citizen, not even a sojourner, but simply a flying traveler? How can he put his whole heart and soul, (without which success in any work is impossible) into living, and doing and being HERE, if all the time his eyes are fixed on an infinitely happier *hereafter?* My *dreaming* may be very beautiful, if it paints the eternal glories of a far-off heavenly state, seen in beatific vision. What must my *doing* be, if it is in accord with the belief that this beautiful world of God's is but a "vale of tears," in which I am "a miserable worm of the dust," doomed to return to the dust and be eaten by other worms?

To this separation of the ideal from the practical in religion, more than to any other cause, in my opinion, do we owe the separation in our modern life between thinking and working; and so between thinkers and workers; a distinction which robs all labor of its rightful honor and dignity. "The Father worketh hitherto and I work," said Jesus. Much of our doing is without thought and much of our thinking without doing to-day. What we want is more thought among our workers and more work among our thinkers,— more of the poet in the street-sweeper, and more of the street-sweeper in the poet, will give us better poets as well as better street sweepers, and in both better MEN. God's lines are not vertical alone; they are also and equally horizontal. They meet at the center at the heart,—and the boundless circumference is their only limit. The immortality of the soul of man, individually, as racially, can only find expression in the unending growth, the perpetual progression, of the body of man, accompanying his soul and one with it in its splendid upward passage from glory unto glory. In the coming *renaissance*, in the millenial Easter morn, whose dawning is already upon us, a larger and livelier faith, based on larger knowledge, fuller light, penetrating humanity in all its members, will find expression in works that shall truly manifest the faith that is in us.

Every true religious movement must be a social, an economic, movement, and every genuine social movement a religious movement. A bond of union between the economic and the religious advance movements of our day is offered in the truth I preach. When men grasp the truth of IMMORTALITY and make it their own, here and now, to be realized on earth and in the flesh, as well as in "Heaven" and in the spirit, they will put their best selves and their whole selves into the work of the world; not, as now, only at infrequent intervals and in widely separated individual instances,—but always and everywhere and all together,—helped, strengthened, cheered and gladdened by true and full recognition of God in man, of eternity here and now, of divinity and immortality in flesh and blood, of conscious recognition and expression of the oneness of God and Man in the universal religion of a brotherhood of humanity!   In this *racial* recognition and manifestation of the Christ, incarnated in humanity, we shall find fulfillment of that prophecy and promise contained in the recognition and manifestation of Christ's incarnation in the individual man Jesus,— without which fuller incarnation the life of Jesus would be meaningless. We shall realize, in that day, the glory to which men have so long looked forward with undying hope:

"That one far off divine event
Towards which the whole creation moves."